W9-CPE-613
Grades 3-5

MATH
masterpieces

Math Skills + Puzzles = Masterpieces

Gunter Schymkiw

Published by World Teachers Press®
www.worldteacherspress.com

Printed in the United States of America.

This book is printed on recycled paper.

Order Number 2-5258
ISBN 978-1-58324-235-3

E F G H I 13 12 11 10 09

395 Main Street
Rowley, MA 01969
www.didax.com

FOREWORD

The activities in *Math Masterpieces* integrate mathematics with aspects of visual arts. As well as being provided with opportunities to consolidate knowledge and skills in mathematics, students are introduced to significant works of art and their artists. *Math Masterpieces* aims to enhance children's appreciation of the works of great artists, at the same time allowing them to build mathematical skills.

Also available: *Math Masterpieces, Grades 6-7*

CONTENTS

Teacher Notes

Each selected artwork is paired with a mathematical concept in a group of three pages.

Page 1 of each group consists of a teacher page. Each teacher page contains the following information.

The **mathematical concept** being covered is indicated at the top of the page.

Background Information includes information about the artist, his/her style, influences and other artworks.

The **Internet Image Search** encourages teachers and students to more broadly appreciate the works of individual artists.

Occasionally, **Additional Activities** are given which may extend the activity into other learning areas.

The **artwork and artist** to which this page relates is given.

A picture of the **completed artwork** shows students the correct answers to their mathematical activity.

A **Talking Mathematically** section is included with each activity. This provides lesson format suggestions, discusses possible mathematical emphases and provides further interesting background information that is both instructional and historical in its nature.

Student Instructions explain how to complete the activity.

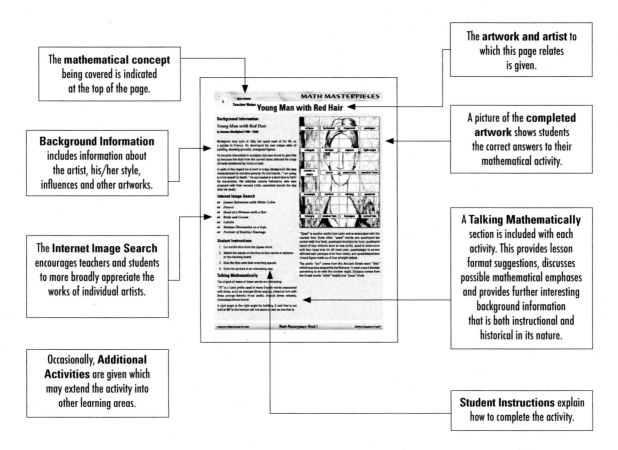

Page 2 of each group consists of the student activity page with the mathematical problems to be solved.

The **artwork** to which this page relates is given.

The **mathematical concept** being covered is indicated at the top of the page.

The **algorithms/problems** for the students to complete are provided.

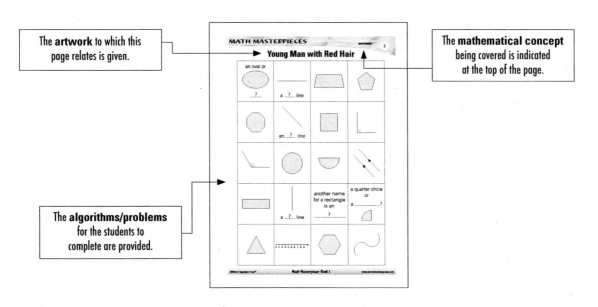

Page 3 of each group provides the answers to the algorithms/problems on "mixed up" sections of the masterpiece.

The **mathematical concept** being covered is indicated at the top of the page.

The **artwork** to which this page relates is given.

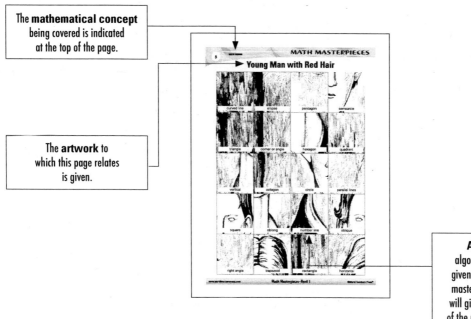

Answers for the algorithms/problems are given on the sections of the masterpiece. These sections will give a completed picture of the masterpiece once glued in the correct position.

NCTM CORRELATION

	Number and Operations	Geometry	Measurement
Young Man with Red Hair		Shapes and Vocabulary	
Color Theory	Addition		Standard Units
Linear Perspective			Time
Still Life with Apples			Calendar
Jonah and the Whale	2-Digit Addition		
Monarch of the Glen	2-Digit Addition		
Ancient Cave Painting	2-Digit Addition		
The Haywain	3-Digit Addition		Money
Henry VIII	2-Digit Addition		
The Sleep of Reason	3-Digit Addition		Money
The Milkmaid	3-Digit Subtraction		
Bubbles	3-Digit Subtraction		Money
Blue Horses	Multiplication		
The Gulf Stream	Multiplication		
The Laughing Cavalier	Division		
Picture from the Bayeux	Division		
Down on His Luck	Division		Money
Starry Night	Fractions		
False Perspective	Multi Operations		

Teacher Notes # Young Man with Red Hair

Background Information

Young Man with Red Hair
by Amedeo Modigliani (1884–1920)

Modigliani was born in Italy but spent most of his life as a painter in France. He developed his own unique style of painting, showing graceful, elongated figures.

He became interested in sculpture but was forced to give this up because the dust from the carved stone affected his lungs (already weakened by tuberculosis).

In spite of the regard he is held in today, Modigliani's life was characterized by extreme poverty. He told friends, "I am going to drink myself to death." He succeeded in a short time to fulfill his declaration. His mistress Jeanne Hebuterne, who was pregnant with their second child, committed suicide the day after his death.

Internet Image Search

☞ *Jeanne Hebuterne with White Collar*

☞ *Pierrot*

☞ *Head of a Woman with a Hat*

☞ *Bride and Groom*

☞ *Lalotte*

☞ *Madam Zborowska on a Sofa*

☞ *Portrait of Beatrice Hastings*

Student Instructions

1. Cut out the tiles from the jigsaw sheet.

2. Match the words on the tiles to their words or pictures on the backing board.

3. Glue the tiles onto their matching spaces.

4. Color the picture in an interesting way.

Talking Mathematically

The origins of many of these words are interesting.

"Tri" is a Latin prefix used in many English words associated with three, such as *triangle* (three angles), *trident* (a fork with three prongs—literally three teeth), *tricycle* (three wheels), *triceratops* (three horns).

A right angle is the right angle for building. A wall that is not built at 90° to the horizon will not stand as well as one that is.

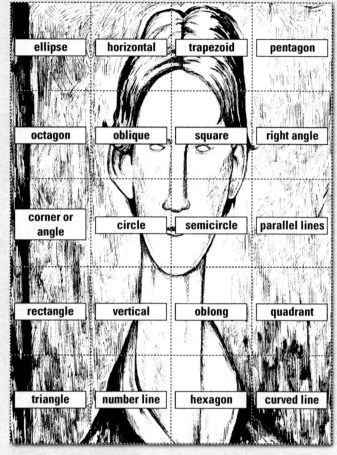

"Quad" is another prefix from Latin and is associated with the number four. Some other *"quad"* words are *quadruped* (an animal with four feet), *quadruple* (multiply by four), *quadruplet* (each of four children born at one birth), *quadriplegic* (a person afflicted with paralysis of all four limbs), and *quadrilateral* (any closed figure made up of four straight sides).

The prefix *"oct"* comes from the Ancient Greek word *"ókto"* which was also adopted by the Romans. In each case, it denotes something to do with the number eight. Octopus comes from the Greek words *"okkto"* (eight) and *"pous"* (foot).

Young Man with Red Hair

an oval or ____?____	a ___?___ line		
	an ___?___ line		
	a ___?___ line	another name for a rectangle is an ___?___	a quarter circle or a _____?
	0 1 2 3 4 5 6 7 8 9		

Young Man with Red Hair

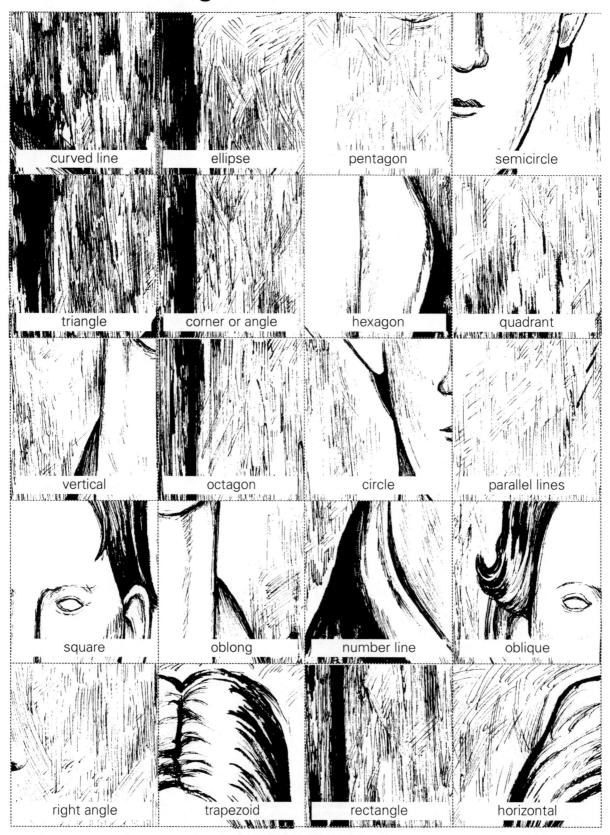

curved line — ellipse — pentagon — semicircle

triangle — corner or angle — hexagon — quadrant

vertical — octagon — circle — parallel lines

square — oblong — number line — oblique

right angle — trapezoid — rectangle — horizontal

Teacher Notes

Color Theory

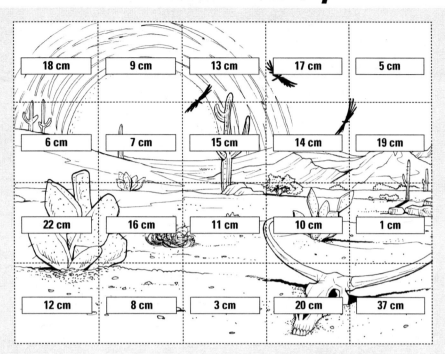

Background Information

Color Theory

Color can be used to create various impressions in art. Red, yellow, orange (all in various shades) and black are often used to create the impression of warmth. The blues, on the other hand, are considered to be cool colors.

Internet Image Search

☞ *Sofala* (by Russell Drysdale)

☞ *The Eagle and the Baobab Tree* (by Clifton Pugh)

☞ *Giraffes* (by Sidney Nolan)

☞ *The MacDonnell Ranges* (by Sidney Nolan)

☞ *Trees on a Hillside* (by Fred Williams)

☞ *The Scream* (by Edvard Munch)

Student Instructions

1. Students add the lengths of the lines in centimeters and write the total.

2. Match the tile numbers with the lengths on the puzzle sheet.

3. Glue the tiles on their correct spaces.

Talking Mathematically

Students may use this method to measure and find the total lengths of shapes made from a number of component lines.

 www.worldteacherspress.com

Color Theory

3.0 cm + 1.5 cm + 0.5 cm	4.5 cm + 4.5 cm + 3.5 cm + 6.5 cm	0.5 cm + 0.5 cm	10.5 cm + 15.5 cm + 11.0 cm
2.5 cm + 3.5 cm + 2.5 cm + 8.5 cm	5.5 cm + 3.5 cm + 3.5 cm + 1.5 cm	3.5 cm + 2.5 cm + 2.0 cm + 2.0 cm	9.5 cm + 4.5 cm + 2.5 cm + 3.5 cm
6.5 cm + 1.5 cm + 1.5 cm + 3.5 cm	9.5 cm + 5.5 cm	2.5 cm + 5.5 cm + 3.0 cm	1.5 cm + 1.5 cm
3.0 cm + 4.5 cm + 1.5 cm	3.5 cm + 2.5 cm + 1.0 cm	6.5 cm + 5.5 cm + 4.0 cm	1.5 cm + 5.5 cm + 1.0 cm
10.5 cm + 1.5 cm + 6.0 cm	2.5 cm + 1.5 cm + 1.5 cm + 0.5 cm	12.5 cm + 4.5 cm + 4.5 cm + 0.5 cm	6.0 cm + 2.0 cm + 2.5 cm + 1.5 cm

Color Theory

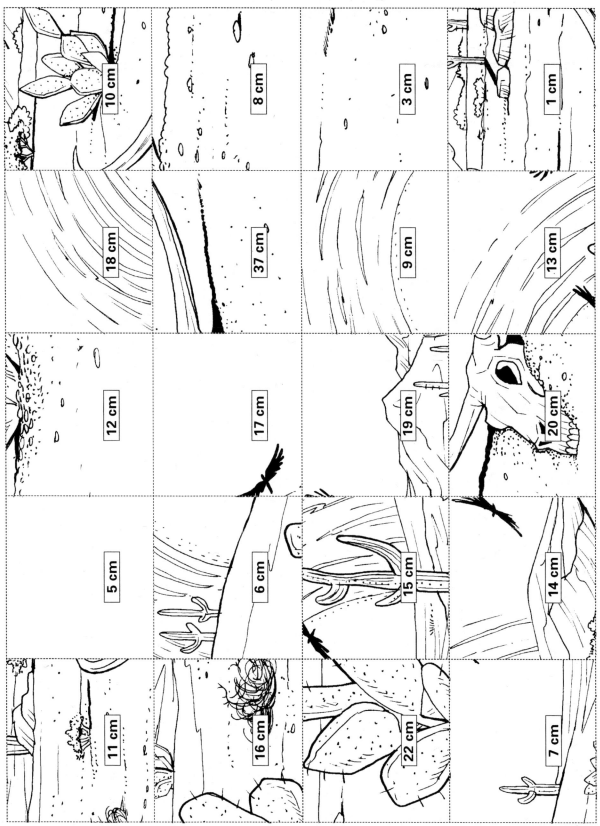

Linear Perspective

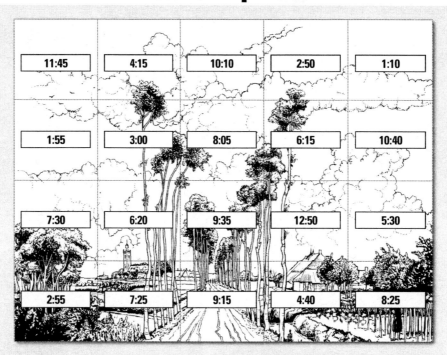

Background Information

Linear Perspective

by Meindert Hobbema (1638–1709)

Perspective is used by artists to give the illusion of depth. Pictures drawn or painted on a flat, two-dimensional surface appear to have three dimensions through the use of perspective.

Linear perspective involves the use of lines converging at a central point (called the vanishing point) on the horizon.

Internet Image Search

- ☞ *Paris Street: A Rainy Day* (by Gustave Caillebotte)
- ☞ *Bourke Street* (by Tom Roberts)
- ☞ *The Last Supper* (by Leonardo da Vinci)
- ☞ *Annunciation* (by Leonardo da Vinci)
- ☞ *Golden Gate Bridge* (by Ray Strong)

Student Instructions

1. Cut out the tiles from the jigsaw sheet.

2. Match each digital time to its analog clock face on the backing board.

3. Glue the tiles onto their matching spaces.

4. Color the picture in an interesting way.

Talking Mathematically

Students can count by fives around the clock face to assist them in becoming familiar with telling the time on an analog clock. When the "to" terminology is used the same technique can be used in lines 3 and 4. For example, when demonstrating 3:40 place a finger on the 8 and count 20, 15, 10 and 5 minutes before the fourth hour. Note that "to" is only used beyond "half past" the hour.

Linear Perspective

Linear Perspective

Still Life with Apples

Teacher Notes

Background Information

Still Life with Apples

by Paul Cezanne (1839–1906)

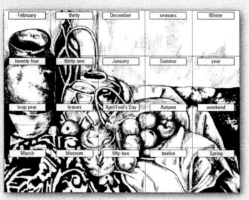

Paul Cezanne was a French artist who experimented with a number of painting styles. He was a leading figure in the art movement known as Impressionism.

Unlike most of the artists who preceded them, Impressionists felt that clarity of detail was not the most important thing in paintings. Impressionists were more interested in the mood created by light and colors in their paintings, so Impressionists' paintings do not have a photographic quality about them. The key subject of the painting does not dominate against a drab background. Rather, Impressionist artists concentrate on their impression of a scene—its colors, shades, light and atmosphere.

Cezanne liked to paint outdoors. In 1906 he was caught in a thunderstorm, became ill, and died a week later.

Internet Image Search

- ☞ *Impression Sunrise (by Claude Monet)*
  (The impressionist movement took its name from this painting)
- ☞ *Les Baigneurs au Repos (Bathers at Rest)*
- ☞ *Landscape at Auvers*
- ☞ *Houses at the Edge of a Road*
- ☞ *Mont St. Victoire*
- ☞ *The Garden at Les Lauves*
- ☞ *Mont de Cengle*
- ☞ *Still Life with Watermelons*

Student Instructions

1. Cut the tiles from jigsaw sheet.
2. Match the words on the tiles to their sentences on the backing board.
3. Glue the tiles onto their matching spaces.
4. Color the picture in an interesting way.

Talking Mathematically

The word *calendar* is linked to the Latin word *kalendarium*, meaning *account book*. This word, however, derives from a word meaning to *call out*. Before there was such a thing as an account book, it was the practice for a herald to proclaim the beginning of each month, and upon this declaration all debts became due for collection. As knowledge of the movement of the earth, moon and stars increased and things could be determined in advance, merchants began keeping their own account books (*kalendaria*). This enabled them to know when debts and interests were due.

The names of the months also derive from Latin and honor important figures, both real and mythological, occasions or, mundanely, their sequence in the year. March was the first month of the Roman calendar and honored *Mars*, the Roman god of war. April is believed to have derived from the Latin word *aperie*, meaning *to open*. Being springtime, this was the time for the opening (blossoming) of trees and flowers. May was named after *Maia* (the goddess of increase), who presided over the growth of plants. June is the month of *Juno*, the Roman queen of heaven. July was the birth month of *Julius Caesar*, the great Roman emperor. It is said that Marc Antony, one of Caesar's murderers, named it after the emperor in the same year that he took part in his assassination. Caesar's nephew, Augustus, followed him as emperor and proclaimed the name August to honor himself. September *(the seventh)*, October *(the eighth)*, November *(the ninth)* and December *(the tenth)* are just named according to their position in the Roman calendar. January honors *Janus*, the god of doors and gates. His head has two faces, one which could look back at the past (the old year) and forward (into the new year). February takes its name from the Latin word *februare*, meaning to *cleanse*. This was the month in which a *cleansing* festival was held.

When the calendar (known as the Julian calendar after Julius Caesar) was revised in 1582 (to the Gregorian calendar after Pope Gregory XIII), the seventh, eighth, ninth and tenth months retained their titles even though their positions were now changed.

MATH MASTERPIECES

Still Life with Apples

December, January and February are the months of ☐☐☐☐ .

There are 365 days in a ☐☐ .

Saturday and Sunday make up the ☐☐☐ .

March, April and May make up the season called ☐☐☐☐ .

Spring, summer, autumn and winter are the ☐☐☐ .

June, July and August are the months of ☐☐☐☐ .

September, October and November are the months of ☐☐☐ .

There are ☐☐☐ months in a year.

Hanukkah is in the month of ☐☐☐☐ .

New Year's Day is on the first of ☐☐☐ .

The first of April is called ☐☐☐☐ .

A year is made up of ☐☐ - ☐☐ weeks.

September, April, June and November all have ☐☐☐ days.

January, March, May, July, August, October and December all have ☐☐ - ☐☐ ☐☐ days.

Some trees lose their ☐☐☐ in autumn.

Many trees ☐☐☐ in spring.

The shortest month is ☐☐☐☐ .

A day has ☐☐ - ☐☐ ☐☐ hours.

There are 366 days in a ☐☐☐ .

The third month is ☐☐☐ .

Still Life with Apples

January
summer
year
April Fool's Day

leaves
weekend
autumn
spring

twenty-four
twelve
fifty-two
blossom

February
March
seasons
thirty-one

December
thirty
leap year
winter

Jonah and the Whale

Background Information

Jonah and the Whale

by Ambrogio Bondone Giotto (1266–1337)

Giotto lived in Italy. He was a farmer's son who was discovered drawing sheep by the famous artist Giovanni Cimabue. Cimabue offered to tutor Giotto, but the student's fame grew to the extent that he overshadowed his master.

The painting of Jonah and the whale is part of a fresco on the walls of the Arena Chapel in the Italian city of Padua. A fresco is a painting done quickly on wet plaster so that the colors penetrate the plaster and become fixed when it dries.

A story is told concerning Giotto and Pope Benedict IX. The pope was interested in employing Giotto and sent a messenger to take back an example of his work. When asked this, Giotto dipped his paintbrush in some red paint and with a single sweep of his hand painted a perfect circle.

Internet Image Search

- ☞ *St. Francis Preaching to the Birds*
- ☞ *Madonna and Child*
- ☞ *The Mourning of Christ*
- ☞ *Expulsion of Joachim from the Temple*
- ☞ *The Visitation*
- ☞ *Crucifixion*
- ☞ *Homage of a Simple Man*
- ☞ *Ascension of Christ*

Student Instructions

1. Cut out the tiles from the jigsaw sheet.

2. Do the additional algorithms on the backing board.

3. Match the numbers on the tiles to the answers on the backing board.

4. Glue the tiles onto their matching spaces.

5. Color the picture in an interesting way.

Talking Mathematically

It is important that the students understand our number system and the importance of place value when doing these algorithms. Algorithms allow us to calculate amounts quickly and efficiently. For example, to combine a group of 30 with a group of 33 (as in the algorithm 30+33) without an algorithm , it would be necessary to first draw 30 things, like this:

* *

and then draw 33 things like this:

* *

and then combine them into a single group like this:

* *
* *

Jonah and the Whale

12 + 11	22 + 21	23 + 32	30 + 33
15 + 14	30 + 10	17 + 21	24 + 32
12 + 14	12 + 20	33 + 32	16 + 21
34 + 41	21 + 40	25 + 24	34 + 32
34 + 42	43 + 42	32 + 22	32 + 32

MATH MASTERPIECES

Jonah and the Whale

Monarch of the Glen

Background Information

Monarch of the Glen

by Sir Edwin Landseer (1802–1873)

Landseer was an Englishman who was to become Queen Victoria's favorite artist. He showed talent at a very early age and had an exhibition of his drawings at London's Royal Academy when he was only 13 years old. Landseer's subjects were mostly animals, but he tried to portray them showing various human expressions. The stag in his most famous painting, *Monarch of the Glen*, looks proud and noble.

Landseer's paintings lost favor in the twentieth century because some people were concerned that some of them seemed to emphasize deer hunting.

Internet Image Search

- ☞ *Deer and Deerhounds in a Mountain Torrent*
- ☞ *Dignity and Impudence*
- ☞ *Marmosets on a Pineapple*
- ☞ *Laying Down the Law*
- ☞ *Flood in the Highlands*
- ☞ *The Old Shepherd's Chief Mourner*
- ☞ *Man Proposes – God Disposes*

Additional Activity

Find and write the dictionary meanings of the words "dignity" and "impudence." Explain why Landseer may have chosen to use these words in the title of his painting *Dignity and Impudence.*

Student Instructions

1. Cut out the tiles from the jigsaw sheet.
2. Do the addition algorithms on the backing board.
3. Match the numbers on the tiles to the answers on the backing board.
4. Glue the tiles onto their matching spaces.
5. Color the picture in an interesting way.

Talking Mathematically

To understand regrouping, students need a sound understanding of place value in the Hindu-Arabic number system. Memory joggers can be used to help. In the first example (15 + 37), when adding the ones column (7 + 5) and getting the answer 12, you might suggest that the students write the two ones in the ones column (together with the other "one-year-olds") and carry the big "10-year-old" up with the other big 10-year-olds (the 10s column). This is a method that seems to strike a chord with some children.

There is, of course, an abundance of materials and explanations that can be used to instill this understanding.

Monarch of the Glen

15 + 37	16 + 17	48 + 46	18 + 17
27 + 37	29 + 36	44 + 36	29 + 45
35 + 18	23 + 37	19 + 73	65 + 16
18 + 14	27 + 34	19 + 29	29 + 38
59 + 34	73 + 18	58 + 39	26 + 56

Monarch of the Glen

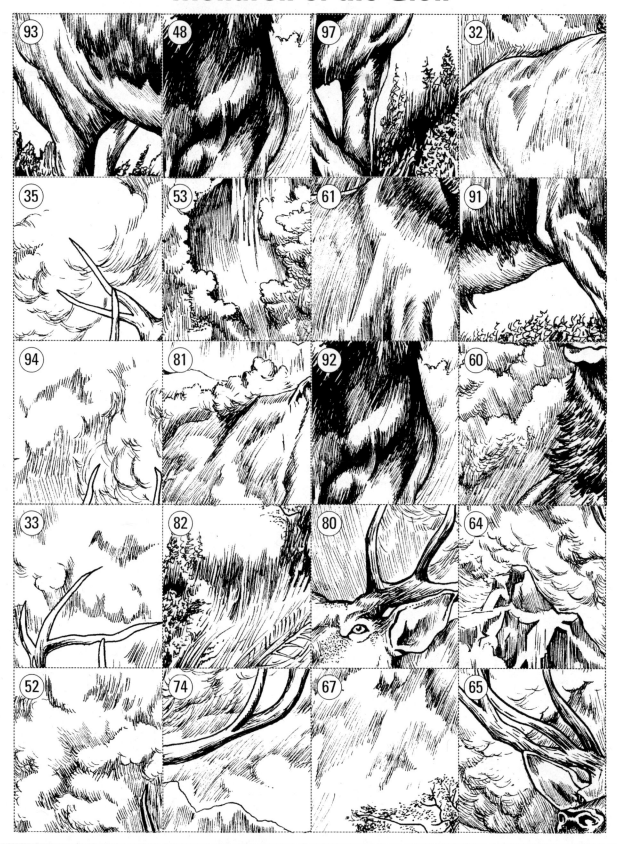

Teacher Notes

Ancient Cave Painting

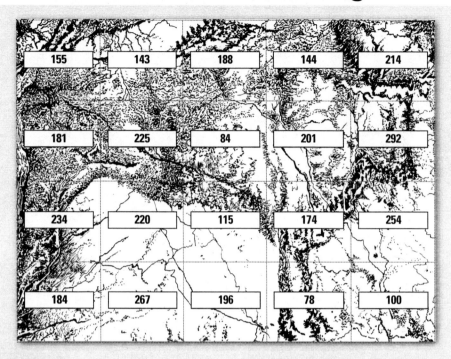

Background Information

Ancient Cave Painting

The cave painting of a bison was painted on a cave wall at Altamira in Spain over 20,000 years ago. Firstly, the artist scratched an outline using something hard, such as flint. After this, the color was filled in using natural materials, such as ground-up minerals. The chief colors of the original painting are brown, red and yellow.

Internet Image Search

☞ *Lascaux Caves*

☞ *Altamira Cave*

Student Instructions

1. Cut out the tiles from the jigsaw sheet.

2. Do the addition algorithms on the backing board.

3. Match the numbers on the tiles to the answers on the backing board.

4. Glue the tiles onto their matching spaces.

5. Color the picture in an interesting way.

Talking Mathematically

There is, of course, an abundance of materials and explanations that can be used to instill this understanding.

Unit column answers on page 25 will sometimes exceed 19. Students who have not fully understood trading will often carry the wrong numbers up to the 10s column. They often become used to carrying one without understanding that it is the number in the units column of the answer that is written. In these examples, students must write both numbers in their answer if the tens column adds to more than nine before attempting the activity if they are not already well-practiced in this type of algorithm.

Ancient Cave Painting

68 64 + 82	97 96 + 99	98 89 + 67	18 66 + 16
49 48 + 47	53 97 + 51	83 87 + 4	33 32 + 13
67 37 + 84	38 17 + 29	88 13 + 14	64 66 + 66
76 18 + 49	65 75 + 85	70 74 + 76	95 95 + 77
32 48 + 75	97 63 + 21	99 99 + 36	63 36 + 85

Ancient Cave Painting

The Haywain

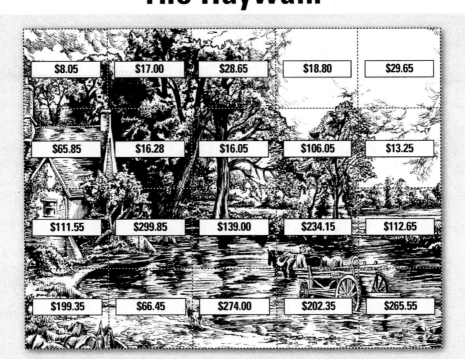

$8.05	$17.00	$28.65	$18.80	$29.65
$65.85	$16.28	$16.05	$106.05	$13.25
$111.55	$299.85	$139.00	$234.15	$112.65
$199.35	$66.45	$274.00	$202.35	$265.55

Background Information

The Haywain

by John Constable (1776–1837)

John Constable was the son of a wealthy merchant and mill owner. Constable loved the countryside, particularly that of his native area of Suffolk in England. Nature always features in his paintings. Clouds, too, are an important part of most of his paintings. *The Haywain* ("wain" is an old word meaning "cart") was awarded a gold medal by the French Emperor, Charles X.

Internet Image Search

☞ *Dedham Church and Vale*

☞ *His Majesty's Ship "Victory"*

☞ *Landscape: Boys Fishing*

☞ *Boat Building*

☞ *Flatford Mill*

Student Instructions

1. Cut out the tiles from the jigsaw sheet.

2. Do the addition algorithms on the backing board.

3. Match the numbers on the tiles to the answers on the backing board.

4. Glue the tiles onto their matching spaces.

5. Color the picture in an interesting way.

Talking Mathematically

Our monetary system clearly demonstrates the decimal system of money. The first column could be described as showing the number of one-cent coins, the second, 10-cent coins and the third, 100-cent (one dollar) coins or bills.

The Haywain

$13.65 1.05 + 14.95	$9.90 1.30 + 2.05	$96.65 13.65 + 2.35	$87.75 78.85 + 98.95
$6.25 3.85 + 8.70	$12.65 14.35 + 79.05	$68.05 78.05 88.05 +	$96.45 45.95 + 59.95
$9.95 9.95 + 8.75	$4.05 2.05 + 9.95	$16.75 84.25 + 38.00	$97.15 98.95 + 77.90
$6.95 1.40 + 8.65	$1.05 9.15 + 6.08	$99.95 99.95 + 99.95	$22.05 11.15 + 33.25
$2.65 1.45 + 3.95	$18.85 19.85 + 27.15	$24.05 17.55 + 69.95	$98.85 67.15 + 33.35

The Haywain

Henry VIII

Background Information

Henry VIII of England

by Hans Holbein (1497–1543)

Holbein was a German painter who moved to England in 1526. He painted portraits of wealthy and powerful people and came to the notice of King Henry VIII, who made him his court painter. When Henry was looking for a new wife (he was married six times), he asked Holbein to travel overseas and paint portraits of likely women. Henry fell in love with Holbein's portrait of Anne of Cleves and married her. Holbein died of bubonic plague in London in 1543.

Internet Image Search

- ☞ *Sir Richard Southwell*
- ☞ *Dorothea Kannengiesser*
- ☞ *Adam and Eve*
- ☞ *The Body of the Dead Christ in the Tomb*
- ☞ *Sir Thomas More*
- ☞ *The Artist's Family*

Student Instructions

1. Cut out the tiles from the jigsaw sheet.
2. Do the subtraction algorithms on the backing board.
3. Match the numbers on the tiles to the answers on the backing board.
4. Glue the tiles onto their matching spaces.
5. Color the picture in an interesting way.

Talking Mathematically

It is important to remind the students that they must always start with the smallest column (i.e. the ones column). This becomes more important when examples involve regrouping.

Henry VIII

50 − 30	69 − 40	84 − 43	75 − 22
67 − 54	87 − 44	98 − 31	99 − 63
58 − 46	79 − 55	87 − 43	68 − 41
96 − 42	97 − 26	66 − 35	88 − 33
96 − 24	79 − 41	89 − 54	68 − 16

Henry VIII

The Sleep of Reason Produces Monsters

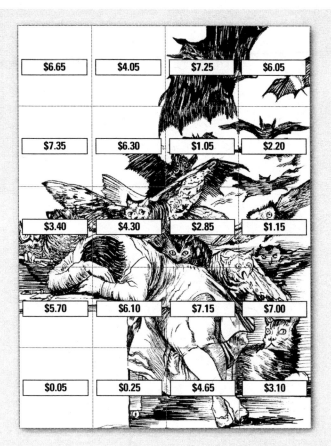

Background Information

The Sleep of Reason Produces Monsters

by Francisco Goya (1746–1828)

The Sleep of Reason Produces Monsters is a self-portrait of the artist at a difficult time in his life. He was stricken by a lengthy illness between 1792 and 1793, leaving him deaf. Goya was Spanish and a very influential painter. He is famous for depicting the horrors of war in a series of paintings showing scenes from Napoleon's invasion and occupation of Spain (1808 – 1814).

Internet Image Search

☞ *The Esquilache Riots*

☞ *The Sacrifice to Vesta*

☞ *The Kite*

☞ *Fight at the Cock Inn*

☞ *The Parasol*

☞ *The Snowstorm*

Student Instructions

1. Cut out the tiles from the jigsaw sheet.

2. Do the subtraction algorithms on the backing board.

3. Match the numbers on the tiles to the answers on the backing board.

4. Glue the tiles onto their matching spaces.

5. Color the picture in an interesting way.

Talking Mathematically

Students should be reminded how our monetary system can be used to demonstrate "what's happening" when doing these subtraction problems; e.g., $7.85 – $1.20.

First of all, we start with the one-cent column: five cents minus nothing leaves us with five single cents in this column. Eight ten-cent coins minus two ten-cent coins leaves six ten-cent coins. Seven one-dollar bills minus one one-dollar bill leaves six one-dollar bills. Totalling these coins and bills give us the answer $6.65.

The Sleep of Reason Produces Monsters

$7.85 – 1.20	$6.05 – 2.00	$8.45 – 1.20	$9.75 – 3.70
$8.35 – 1.00	$9.95 – 3.65	$8.15 – 7.10	$9.90 – 7.70
$4.95 – 1.55	$7.60 – 3.30	$4.95 – 2.10	$7.45 – 6.30
$9.85 – 4.15	$9.30 – 3.20	$9.85 – 2.70	$8.10 – 1.10
$9.05 – 9.00	$8.65 – 8.40	$9.95 – 5.30	$8.90 – 5.80

The Sleep of Reason Produces Monsters

The Milkmaid

Background Information

The Milkmaid

by Johannes Vermeer (1632–1675)

Vermeer was born and lived all of his life in the Dutch town of Delft. Many of his paintings show people doing everyday tasks inside, just as his milkmaid is doing.

Vermeer had 11 children and only produced a small number of paintings each year (only 35 known works exist). When he died unexpectedly at the age of 43, he was heavily in debt.

Internet Image Search

- ☞ *The Love Letter*
- ☞ *Christ in the House of Mary and Martha*
- ☞ *Diana and her Companions*
- ☞ *Girl Asleep at a Table*
- ☞ *The Glass of Wine*
- ☞ *The Music Lesson*
- ☞ *The Geographer*
- ☞ *The Astronomer*

Student Instructions

1. Cut out the tiles from the jigsaw sheet.

2. Do the subtraction algorithms on the backing board.

3. Match the numbers on the tiles to the answers on the backing board.

4. Glue the tiles onto their matching spaces.

5. Color the picture in an interesting way.

Talking Mathematically

Just as with the equal addition of 10s (borrowing and paying back), there is a clear way of explaining what is going on for the equal subtraction of 10s.

For example, for 31 − 19, show 31 as three groups of 10 and one (unit).

Draw boundary ropes around each bundle of 10.

Ask the students: Have I added any objects? (No)

Have I taken any objects away? (No)

What I have done to organize the objects differently. (Instead of three bundles of 10 and a single one, I have organized them into two bundles of 10 and 11 ones.)

The Milkmaid

31 − 19	90 − 17	42 − 19	70 − 19
61 − 48	71 − 39	60 − 19	71 − 49
80 − 16	72 − 39	70 − 55	74 − 37
51 − 32	80 − 28	52 − 26	81 − 37
90 − 19	34 − 17	91 − 19	50 − 16

The Milkmaid

Bubbles

Background Information

Bubbles

by Sir John Everett Millais (1829–1896)

Millais was an Englishman. At the age of 11 he was the youngest artist ever to join the Royal Academy, Britain's finest art school. He paid great attention to detail in his paintings, and his background landscapes were painted from nature. Millais and his wife had eight children. His paintings were very popular and sold at high prices during his lifetime. *Bubbles* is a painting of his grandson, Willie. It was used by a soap company in an advertisement for its product.

Internet Image Search

☞ *The Bridesmaid*

☞ *The Return of the Dove to the Ark*

☞ *The Blind Girl*

☞ *Apple Blossoms*

☞ *The Boyhood of Raleigh*

Student Instructions

1. Cut out the tiles from the jigsaw sheet.

2. Do the subtraction algorithms on the backing board.

3. Match the number on the tiles to the answers on the backing board.

4. Glue the tiles onto their matching spaces.

5. Color the picture in an interesting way.

Talking Mathematically

Money Tales

#1 Why are piggy banks so-called?

In the 1500s a type of clay, commonly used in the making of utensils such as pots, jars and dishes was called pygg. Many households put aside a pot or jar made of pygg for spare coins. Over time the origin of the name of this special home bank was forgotten, and it evolved to take on the name piggy. Creative potters began making such containers in the shape of a pig.

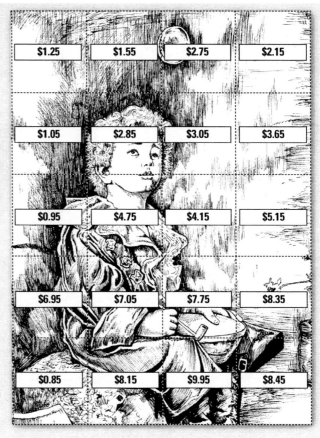

$1.25	$1.55	$2.75	$2.15
$1.05	$2.85	$3.05	$3.65
$0.95	$4.75	$4.15	$5.15
$6.95	$7.05	$7.75	$8.35
$0.85	$8.15	$9.95	$8.45

#2 Why do our coins have milled (grooved) edges?

In earlier times coins were made of precious metal. It was a common practice among some people (notably merchants who handled many coins in a day's business) to clip the edges of coins, and after accumulating a reasonable amount of the precious metal, melt it and sell it. During the reign of the English king, Henry V (1413 – 1422) the practice became so widespread that the death penalty was applied to anyone found guilty of doing it. In spite of this, some people could not resist the lure of "easy money" and new measures had to be taken. A milled rim around the edge of coins was introduced. Any coin with part of this missing was deemed to not be legal tender (currency that someone must accept as payment for a debt).

MATH MASTERPIECES

Bubbles

$3.00 − 1.75	$6.30 − 4.75	$9.65 − 6.90	$9.00 − 6.85
$6.10 − 5.05	$5.15 − 2.30	$6.75 − 3.70	$7.30 − 3.65
$6.00 − 5.05	$9.50 − 4.75	$8.05 − 3.90	$9.75 − 4.60
$9.25 − 2.30	$8.05 − 1.00	$9.95 − 2.20	$8.60 − 0.25
$6.95 − 6.10	$9.60 − 1.45	$10.00 − 0.05	$9.65 − 1.20

Bubbles

Teacher Notes

Blue Horses

Background Information

Blue Horses

by Franz Marc (1880–1916)

Franz Marc was a German painter who preferred to paint animals because he felt they lived in true harmony with nature. Marc was influenced by the art movement known as Expressionism. Expressionist painters emphasize emotions rather than realism in their paintings. A compelling example of an Expressionist painting is Edvard Munch's *The Scream*. Although not a realistic representation of a person, the emotion of terror is clearly conveyed.

Marc was made to join the German army during WWI. He was killed in action at Verdun in France in 1916.

426	639	648	960	468
862	396	996	860	406
828	848	884	369	822
486	806	2,448	4,862	9,639

Internet Image Search

☞ *Bewitched Mill*

☞ *Indersdorf*

☞ *Two Cats, Blue and Yellow*

☞ *Three Cats*

☞ *Tiger*

☞ *Foxes*

Student Instructions

1. Cut out the tiles from the jigsaw sheet.

2. Do the multiplication algorithms on the backing board.

3. Match the numbers on the tiles to the answers on the backing board.

4. Glue the tiles onto their matching spaces.

5. Color the picture in an interesting way.

Talking Mathematically

It is important to understand why we use algorithms. In general it has to do with efficiency. By writing a few "squiggles" (numerals) and applying an algorithm correctly we can calculate large amounts which would take a great deal of time and effort to calculate by counting.

Multiplication is a shortened method of addition of equal amounts.

Using the algorithm

$$\begin{array}{r} 3{,}213 \\ \times \quad 3 \\ \hline \end{array}$$

as an example, ask students how an answer could be found without using an algorithm.

To count three groups of 3,213 would take a great deal of time and effort and is more likely to be inaccurate due to factors such as losing count, being interrupted, etc.

As an addition algorithm, an answer can be achieved by writing 12 numerals and ruling one line.

$$\begin{array}{r} 3{,}213 \\ 3{,}213 \\ +\ 3{,}213 \\ \hline \end{array}$$

Using a multiplication algorithm, we can shorten this to five numerals and one line.

$$\begin{array}{r} 3{,}213 \\ \times \quad 3 \\ \hline \end{array}$$

Blue Horses

234 × 2	320 × 3	324 × 2	213 × 3
203 × 2	430 × 2	332 × 3	132 × 3
411 × 2	123 × 3	442 × 2	424 × 2
3,213 × 3	2,431 × 2	1,224 × 2	403 × 2
		213 × 2	431 × 2
		414 × 2	243 × 2

MATH MASTERPIECES

Blue Horses

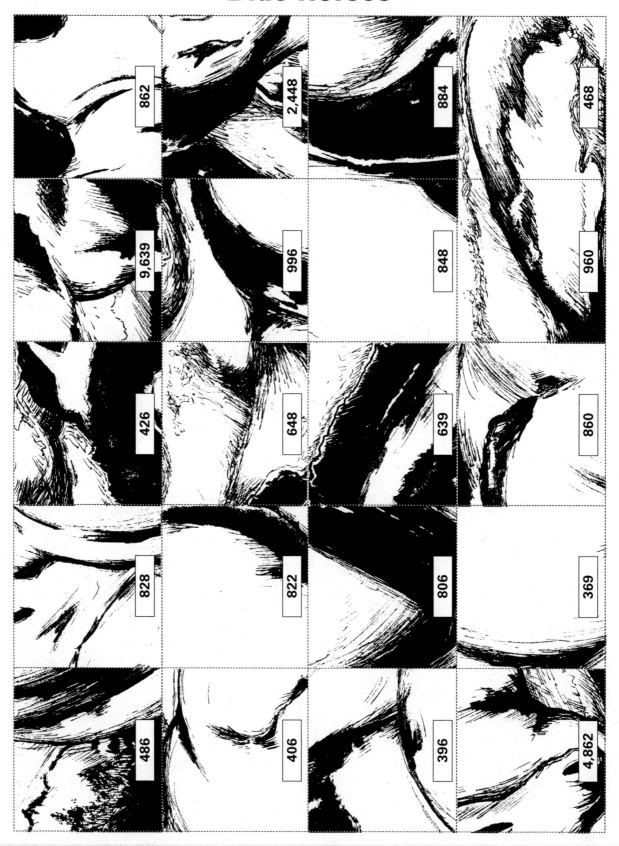

The Gulf Stream

Teacher Notes

Background Information

The Gulf Stream

by Winslow Homer (1836–1910)

Homer was an American painter who lived at various times in France, England, Bermuda and the Bahamas. He worked as an illustrator and war artist for the magazine *Harper's Weekly* during the Civil War. Many of his paintings are on the theme of people setting themselves against nature. In *The Gulf Stream*, we see the man in a schooner whose mast has been snapped. A waterspout is approaching, and the shark-infested water offers no escape.

Internet Image Search

☞ *Snap the Whip*

☞ *Eight Bells*

☞ *The Sharpshooter*

☞ *The Herring Net*

☞ *The Fog Warning*

☞ *Right and Left*

☞ *Sponge Fishermen*

☞ *The Bridle Path*

☞ *Kissing the Moon*

Additional Activities

Discuss the questions:

- Which of these pictures was an illustration?
- What are the children doing in *Snap the Whip*?
- Why do you think the painting has this name?

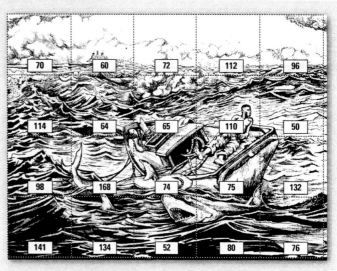

Student Instructions

1. Cut out the tiles from the jigsaw sheet.

2. Do the multiplication algorithms on the backing board.

3. Match the numbers on the tiles to the answers on the backing board.

4. Glue the tiles onto their matching spaces.

5. Color the picture in an interesting way.

Talking Mathematically

Demonstrate how repeated addition of equal amounts mimics the processes involved in multiplication with regrouping.

$$\begin{array}{r} 1\,6 \\ \times\,5 \\ \hline \textbf{80} \end{array}$$

can also be calculated using the algorithm

$$\begin{array}{r} {}^{3} \\ +1\,6 \\ +1\,6 \\ +1\,6 \\ +1\,6 \\ +1\,6 \\ \hline \textbf{80} \end{array}$$

The total of the ones column can be calculated by adding 6 five times:

$6 + 6 + 6 + 6 + 6 = 30.$

Another way of saying this is five groups of 6. Using the addition algorithm we write down the zero and "carry" the 3 to the 10s column (to be added later). Using the multiplication algorithm, we write down the zero and "carry" the 3 to the 10s column (also to be added later). To conclude the addition algorithm, we add five groups of one and then add the amount carried.

MATH MASTERPIECES

The Gulf Stream

48 × 2 ___	25 × 2 ___	66 × 2 ___	38 × 2 ___
56 × 2 ___	55 × 2 ___	15 × 5 ___	16 × 5 ___
36 × 2 ___	13 × 5 ___	37 × 2 ___	26 × 2 ___
15 × 4 ___	16 × 4 ___	56 × 3 ___	67 × 2 ___
35 × 2 ___	57 × 2 ___	49 × 2 ___	47 × 3 ___

The Gulf Stream

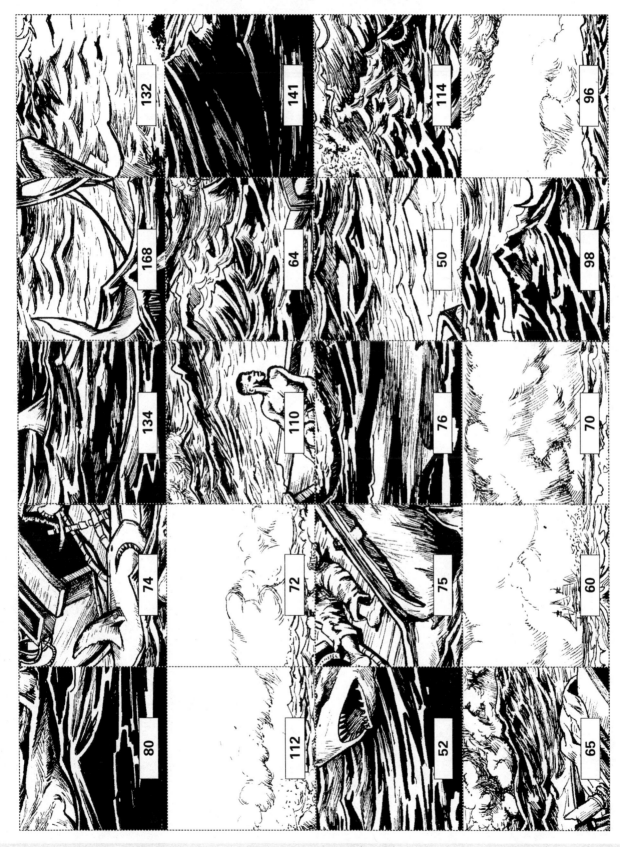

The Laughing Cavalier

Background Information

The Laughing Cavalier

by Frans Hals (c. 1580–1666)

Hals was a Dutch portrait painter who was able to capture the character of his subjects more than any artist before his time. Despite his great skill and the popularity of his work, he was penniless when he died.

Internet Image Search

☞ *The Governors of the Old Men's Almhouse at Haarlem*

☞ *Boy with a Flute*

☞ *Jester with a Lute*

Student Instructions

1. Cut out the tiles from the jigsaw sheet.

2. Do the division algorithms on the backing board.

3. Match the numbers on the tiles to the answers on the backing board.

4. Glue the tiles onto their matching spaces.

5. Color the picture in an interesting way.

Talking Mathematically

Various approaches can be used to demonstrate division. For example, 26 "divided by/shared between" 2 can be demonstrated by sharing 26 items between two students.

Draw 26 "sticks of chalk."

As you give one each to "Bruce" and "Raelene" (or whomever you choose), circle and label these on your diagram. For example:

$$\text{(I)}^B \ \text{(I)}^R \ \text{(I)}^B \ \text{(I)}^R \ \text{(I)}^B \ \text{(I)}^R \ \text{(I)}^B \ \text{(I)}^R \ \text{(I)}^B \ \text{(I)}^R \ \text{(I)}^B \ \text{(I)}^R$$

$$\text{(I)}^B \ \text{(I)}^R \ \text{(I)}^B \ \text{(I)}^R \ \text{(I)}^B \ \text{(I)}^R \ \text{(I)}^B \ \text{(I)}^R \ \text{(I)}^B \ \text{(I)}^R \ \text{(I)}^B \ \text{(I)}^R$$

$$\text{(I)}^B \ \text{(I)}^R$$

Recipient of each piece (B = Bruce, R = Raelene in this example)

Ask each student how many pieces of chalk they have.

Ask students how many times B was written (for Bruce) and how many times R was written (for Raelene).

Using the short division algorithm $2\overline{)26}$, you are individually calculating "How many groups of 10 will each child get if they share two groups of 10?" and "How many ones will each child get if they share six ones?"

The Laughing Cavalier

$2\overline{)22}$	$2\overline{)44}$	$2\overline{)66}$	$2\overline{)24}$
$2\overline{)46}$	$2\overline{)80}$	$2\overline{)60}$	$2\overline{)68}$
$2\overline{)48}$	$2\overline{)82}$	$2\overline{)26}$	$2\overline{)42}$
$2\overline{)62}$	$2\overline{)84}$	$2\overline{)28}$	$2\overline{)64}$
$2\overline{)86}$	$2\overline{)88}$	$2\overline{)222}$	$2\overline{)240}$

MATH MASTERPIECES

The Laughing Cavalier

Picture from the Bayeux Tapestry

Background Information

Picture from the Bayeux Tapestry

The Bayeux Tapestry is an embroidered cloth about 230 feet long and 1.6 feet wide. It illustrates events leading up to and including the Battle of Hastings in 1066. It was this battle between the Norman army led by William the Conqueror and the Saxons led by Harold, Earl of Wessex, that ensured the success of the Norman invasion of England. It took eleven years (from 1066 to 1077) to make the tapestry. It is not really known who made the tapestry. Some historians credit Matilda, William's wife, with making it. Others believe it was stitched by Englishmen from Canterbury.

Internet Image Search

☞ *Bayeux Tapestry*

Student Instructions

1. Cut out the tiles from the jigsaw sheet.

2. Do the division algorithms on the backing board.

3. Match the numbers on the tiles to the answers on the backing board.

4. Glue the tiles onto their matching spaces.

5. Color the picture in an interesting way.

Talking Mathematically

To assist with these algorithms initially, diagrammatic "crutches" may be provided to assist students with their calculations. For example:

$$4\overline{)99}$$

Diagram: tens / / / / / / / / / ones / / / / / / / / /

Use the diagram to demonstrate nine (10s) divided by 4 (circle two groups of 4).

This leaves one group of 10 and nine ones (19).

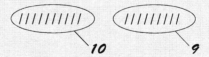

Divide 19 by 4 (circle 4 groups of 4).

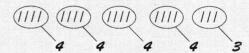

Three things "remain" left over.

We call these the "remainder."

Picture from the Bayeux Tapestry

$2\overline{)35}$	$5\overline{)74}$	$5\overline{)87}$	$4\overline{)99}$
$2\overline{)55}$	$5\overline{)84}$	$4\overline{)53}$	$2\overline{)71}$
$2\overline{)53}$	$2\overline{)75}$	$2\overline{)37}$	$2\overline{)91}$
$2\overline{)51}$	$4\overline{)57}$	$4\overline{)87}$	$4\overline{)55}$
$5\overline{)69}$	$2\overline{)77}$	$5\overline{)89}$	$2\overline{)95}$

Picture from the Bayeux Tapestry

54

Teacher Notes

Down on His Luck

Background Information

Down on His Luck

by Frederick McCubbin (1855–1917)

Frederick McCubbin was the son of a Melbourne baker. Recognizing his son's talent, his father helped him get a job with a coachmaker where he painted decorations and crests on the coaches. McCubbin was unequaled at painting the Australian bush. His paintings usually had a story to tell, often of the hardships of rural life. This is shown in the expression and setting of *Down on His Luck*.

Internet Image Search

☞ *The Lost Child*

☞ *A Bush Burial*

☞ *On the Wallaby Track*

☞ *The Pioneer*

☞ *Winter Sunlight*

Student Instructions

1. Cut out the tiles from the jigsaw sheet.

2. Do the division algorithms on the backing board.

3. Match the numbers on the tiles to the answers on the backing board.

4. Glue the tiles onto their matching spaces.

5. Color the picture in an interesting way.

Talking Mathematically

More Money Tales

What is the origin of the word "dollar"?

A *thaler* (pronounced tar-lar) was a coin made from metal mined in *Joachimsthal* (Joachim's Valley) in the Czech Republic. People mispronounced this "dol-lar." The term became a general name for currency. The dollar is now the name given to the official currency of 37 countries.

Down on His Luck

$2\overline{)\$2.50}$	$5\overline{)\$4.95}$	$3\overline{)\$8.67}$	$4\overline{)\$9.96}$
$4\overline{)\$1.64}$	$2\overline{)\$4.40}$	$4\overline{)\$4.56}$	$2\overline{)\$9.98}$
$2\overline{)\$1.78}$	$2\overline{)\$1.34}$	$2\overline{)\$5.08}$	$5\overline{)\$8.65}$
$2\overline{)\$0.04}$	$4\overline{)\$4.64}$	$4\overline{)\$4.12}$	$2\overline{)\$6.04}$
$2\overline{)\$1.36}$	$5\overline{)\$1.05}$	$3\overline{)\$7.38}$	$2\overline{)\$3.44}$

MATH MASTERPIECES

Down on His Luck

Starry Night

Background Information

Starry Night

by Vincent van Gogh (1853–1890)

Vincent van Gogh was a brilliant artist but led an unsettled life because of recurring periods of mental disturbance. He only began painting at the age of 27, but in the ten-year period of activity before he shot and killed himself, he completed 1,600 paintings.

He was not recognized until after his death, and most of his life was lived in abject poverty. He was a friend of the French artist Paul Gauguin, and it was after an argument with him that van Gogh cut off part of his own ear. Most of his paintings are made up of separate unblended brush marks. This technique is clearly seen in *Starry Night.*

Van Gogh always signed his pictures with his first name, Vincent, only.

Internet Image Search

☞ *Sunflowers*

☞ *Seashore at Scheveningen*

Student Instructions

1. Cut out the tiles from the jigsaw sheet.
2. Write the fraction indicated by the shading on each shape on the backing board.
3. Match the fractions on the tiles to the fractions on the backing board.
4. Glue the tiles onto their matching spaces.
5. Color the picture in an interesting way.

Talking Mathematically

It is important that students clearly understand what the numerator and denominator of a fraction stand for. Students can often tell you the names numerator and denominator but are not always clear on their meaning.

The **denominator** (bottom number) tells us the number of equal-sized parts a shape is divided (cut, etc.) into.

The **numerator** is the number of these parts that we are considering (talking about).

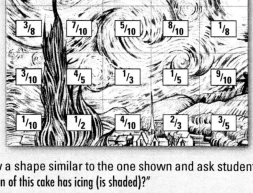

Draw a shape similar to the one shown and ask students, "What fraction of this cake has icing (is shaded)?"

To find the answers, follow these steps.

Step 1 – *Draw the fraction bar (line separating numerator and denominator).*

Step 2 – *The denominator (bottom number) tells us the number of equal-sized parts a shape is divided (cut, etc.) into. This cake is divided into five equal-sized pieces, so the bottom number is 5.*

Step 3 – *The numerator is the number of these parts that we are talking about.*

The question asked is, "What fraction of this cake has icing (is shaded)?" We are talking about the shaded parts. There are two parts shaded so the top number is 2.

The fraction then is $^2/_5$.

Now ask students, "What fraction of this cake does not have icing (is not shaded)?"

Follow these steps.

Step 1 – *Draw the fraction bar (line separating the numerator and denominator).*

Step 2 – *The denominator (bottom number) tells us the number of equal-sized parts a shape is divided (cut, etc.) into. This cake is divided into five equal sized pieces, so the bottom number is 5.*

Step 3 – *The numerator is the number of these parts that we are considering (talking about).*

The question asked is, "What fraction of this cake does not have icing (is not shaded)?"

We are talking about the unshaded parts. There are three parts that are not shaded so the top number is 3.

The fraction then is $^3/_5$.

Starry Night

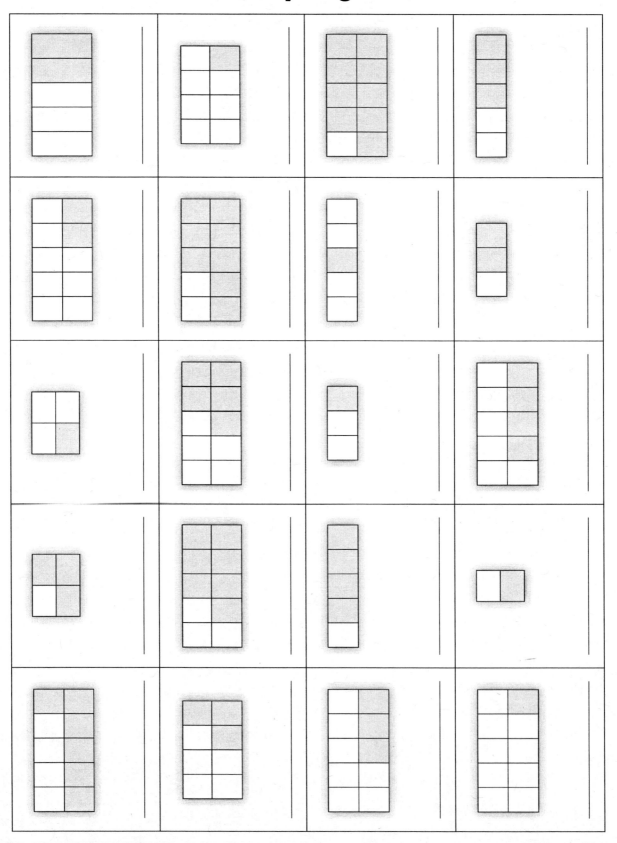

Starry Night

False Perspective

Background Information

False Perspective

by William Hogarth (1697–1764)

Hogarth was an English artist whose career in art began as a book illustrator in 1720. He studied painting in his spare time. Hogarth was an entrepreneur and chose topics and themes that he knew would be popular with the art-buying public.

His paintings often show things that were topical in his time. Many of his paintings satirize people's foibles.

Hogarth's purpose in creating the engraving *False Perspective* was to show how absurd it was for someone to attempt becoming an artist without having an understanding of perspective.

Internet Image Search

☞ *The Fishing Party*

☞ *The Theft of a Watch*

☞ *The Tavern Scene*

☞ *Sarah Malcom in Prison*

☞ *The Good Samaritan*

☞ *The Shrimp Girl*

☞ *Gin Lane*

☞ *Beer Street*

☞ *Shortly After the Marriage*

Student Instructions

1. Cut out the tiles from the jigsaw sheet.

2. Write the number indicated by each description on the backing board.

3. Match the numbers on the tiles to the answers on the backing board.

4. Glue the tiles onto the matching spaces.

5. Color the picture in an interesting way.

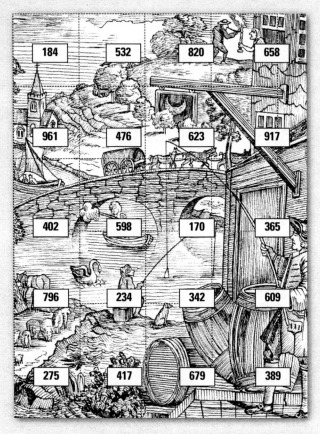

Talking Mathematically

The Hindu-Arabic system (also called the decimal system) was popularized in Europe by the Italian mathematician Leonardo Fibonacci (1170 – 1250), who was also known as Leonardo of Pisa. Until that time, Roman numerals were used.

The decimal system uses ten digits (0 1 2 3 4 5 6 7 8 9) to write all numbers. Its basis is a system of place value where each column has a value that is ten times that immediately to its right. This holds true even with decimal fractions (decimals), where any number to the right of the decimal point has one-tenth of its value in the ones column.

Unlike most number systems, the decimal system has a numeral for zero. Zero is used as a place holder in larger numbers.

False Perspective

4 ones 8 tens 1 hundred _____	3 tens 5 hundreds 2 ones _____	0 ones 2 tens 8 hundreds _____	5 tens 6 hundreds 8 ones _____
60 + 1 + 900 _____	400 + 6 + 70 _____	3 + 600 + 20 _____	900 + 7 + 10 _____
(4 x 100) + (2 x 1) _____	(5 x 100) + (9 x 10) + (8 x 1) _____	(1 x 100) + (7 x 10) _____	(3 x 100) + (6 x 10) + (5 x 1) _____
700 90 + 6 _____	200 30 + 4 _____	300 40 + 2 _____	600 + 9 _____
7 tens + 5 ones + 2 hundreds _____	10 + 400 + 7 _____	(6 x 100) + (7 x 10) + (9 x 1) _____	300 80 + 9 _____

False Perspective

Addition Patterns

1

Complete this table of addition facts

+	0	1	2	3	4	5	6	7	8	9
0										
1										
2										
3										
4										
5										
6										
7										
8										
9										

= _____ 1st row

= _____ 2nd row

= _____ 3rd row

= _____ 4th row

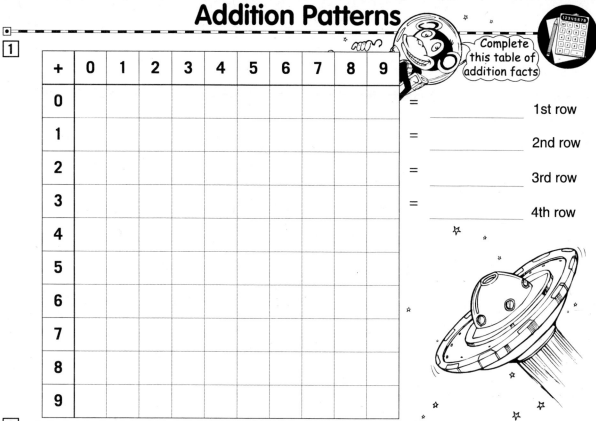

2 (a) Add all of the numbers in the first row, second row, third row and fourth row.

(b) Write down any patterns that you notice. _____

_____

3 (a) Predict the results of adding the numbers in the

(i) 5th row _____

(ii) 6th row _____

(iii) 7th row _____

(iv) 8th row _____

(b) Add the numbers in each row to check your predictions.

Check the answer if you predicted correctly.

I could use the answer from adding the numbers in the first row to help me work this out.

4 Predict the results for adding …

(a) $10 + 11 + 12 + 13 + 14 + 15 + 16 + 17 + 18 + 19 =$ _____

(b) $12 + 13 + 14 + 15 + 16 + 17 + 18 + 19 + 20 + 21 =$ _____

(c) $10 + 20 + 30 + 40 + 50 + 60 + 70 + 80 + 90 =$ _____

Noticing Nines–1

There are many different patterns that may be found in the multiplication tables.
Consider some from the nine times table.

1. (a) Complete the nine times table.

1 x 9 = _____	
2 x 9 = _____	
3 x 9 = _____	
4 x 9 = _____	
5 x 9 = _____	
6 x 9 = _____	
7 x 9 = _____	
8 x 9 = _____	
9 x 9 = _____	
10 x 9 = _____	

What happens when you add the digits?
e.g. 2 x 9 = 18
1 + 8 = 9

_____ 9 _____ = _____ 9
_____ 1 + 8 _____ = _____ 9
_____ = _____
_____ = _____
_____ = _____
_____ = _____
_____ = _____
_____ = _____
_____ = _____
_____ = _____

What do you notice about the tens?

What do you notice about the ones?

What do you notice about the answers?

(b) Record your observations. _____

_____

_____

2. What happens if, instead of adding the digits in the answer, you subtract the smaller number from the larger? e.g. (4 x 9 = 36, 6 – 3 = 3.)
Describe the pattern that is formed.

_____

_____

_____

_____

Investigate here.